Nora the Naturalist's Animals

Farm Animals

A+
Smart Apple Media

Published by Smart Apple Media, an imprint of Black Rabbit Books
P.O. Box 3263, Mankato, Minnesota 56002
www.blackrabbitbooks.com

U.S. publication copyright © 2014 Smart Apple Media. International copyright reserved in all countries.
No part of this book may be reproduced in any form without written permission from the publisher.

Produced by David West 👫 Children's Books
7 Princeton Court, 55 Felsham Road, London SW15 1AZ

Designed and illustrated by David West

Copyright © 2013 David West Children's Books

Library of Congress Cataloging-in-Publication Data

West, David, 1956-
  Farm animals / David West.
    p. cm. – (Nora the Naturalist's animals)
  Summary: "Nora the Naturalist discovers the world of farm animals, describing various animals and how they live."– Provided by publisher.
  Includes index.
  ISBN 978-1-62588-000-0 (library binding)
  ISBN 978-1-62588-049-9 (paperback)
1. Domestic animals–Juvenile literature. 2. Livestock–Juvenile literature. I. Title.
SF75.5.W456 2014
636–dc23

2013007136

Printed in China
CPSIA compliance information: DWCB13CP
010313

9 8 7 6 5 4 3 2 1

Nora the Naturalist says:
I will tell you something more about the animal.

Learn what this animal eats.

Where in the world is the animal found?

Its size is revealed!

What animal group is it – mammal, bird, reptile, amphibian, insect, or something else?

Interesting facts.

Contents

Cattle 4
Sheep 6
Pigs 8
Chickens 10
Goats 12
Ducks 14
Geese 16
Horses 18
Donkeys and Mules 20
Cats and Dogs 22
Glossary and Index 24

Cattle are herbivores. That means they only eat plants.

Cattle live all over the world and on every continent except Antarctica.

An average cow grows to around 900 to 1,100 lbs (408 to 498 kilos).

Cattle are a type of mammal that are called ruminants. Ruminants chew and swallow in the normal way, then regurgitate the semi-digested cud and chew it again to extract its maximum possible food value.

There are over 800 different breeds of cattle in the world.

Bull

Nora the Naturalist says:
Male cattle are called bulls, female cattle are called cows, and baby cattle are called calves.

Cattle

On ranches and farms around the world large herds of cattle are kept for producing milk. Herds of **steers** or **bullocks** are also farmed to supply beef.

Cow

Calf

Sheep

Sheep were one of the first animals to be **domesticated** for farming purposes. Sheep are raised for their fleece, meat, and milk.

Lamb

Ram

 Sheep are herbivores.

 Sheep live all over the world and on every continent except Antarctica.

 An average sheep grows to around 100-150 pounds (45 to 68 kilos).

 Sheep are members of the bovid family – cloven-hoofed, ruminant mammals.

 There are over one billion sheep on Earth – more than any other species of animal.

Ewe

Nora the Naturalist says: Male sheep are called rams, female sheep are called ewes and baby sheep are called lambs.

7

Pigs

Piglets

There are about one billion pigs on farms around the world today. They are commonly raised by farmers for their meat, generally called pork, ham, and bacon.

Nora the Naturalist says:
Male pigs are called boars, female pigs are called sows, and baby pigs are called piglets.

Boar

Sow

 Pigs are omnivores, which means that they eat both plants and animals.

 Pigs live all over the world and on every continent except Antarctica.

 Pigs can grow up to 71 inches (1.8 m) and weigh up to 770 pounds (350 kg). That's the same weight as three humans.

 Pigs are hoofed animals that belong to the mammal group.

 Pigs are also farmed for leather, and their bristly hairs are used to make brushes.

Chickens in the wild eat plants and meat. They scratch at the soil to find seeds, insects, and even lizards or small mice.

Chickens are one of the most common and widespread domestic animals and are found all over the world.

Chickens grow up to 16 inches (0.4 m) and usually weigh around 6 pounds (2.7 kg).

Chickens are birds. Although they can fly they prefer not to because they are too heavy.

Some breeds of hen can produce over 300 eggs each in a year. Roosters often make a loud call at sunrise, that sounds like "cock-a-doodle-do."

Roosters

Nora the Naturalist says:
Male chickens are called roosters, female chickens are called hens, and baby chickens are called chicks.

Chickens

Chickens farmed for meat are called broiler chickens. Chickens farmed for eggs are called egg-laying hens.

Hens

Chicks

Goats are herbivores although people think they will eat almost anything. They usually browse on the tips of woody shrubs and trees.

Goats can be found all over the world. They are very intelligent and can be trained to pull carts.

Average goats grow to around 25 inches (0.64 m) in height and weigh around 100 pounds (45 kg).

Goats are mammals and are closely related to sheep. Some breeds grow soft coats, like the Angora, so are kept for their wool.

Goats are one of the earliest domesticated animals. There are over 300 different breeds of goat.

Billy goat

Nanny goat

Goats

Goats are farmed for their milk, meat, hair, and skins all over the world. In some places people keep them as pets.

Kid

Nora the Naturalist says:
Female goats are called nannies, males are called billies, and baby goats are kids.

Drake   Duck

Ducks eat a variety of food such as grasses, aquatic plants, fish, insects, small amphibians, and worms.

Ducks are farmed all over the world. They are most popular in China and South East Asia.

Average ducks grow to around 16 inches (0.4 m) in length and weigh around 3 pounds (1.4 kg).

Ducks are members of the bird family. Domesticated ducks have their wings clipped so that they cannot fly away.

Ducks talk to each other by "quacking."

Ducks

Ducks spend most of their time on water. They are farmed for their meat, eggs, and soft feathers. Most farmed ducks are white and are descended from the Mallard duck.

Ducklings

Nora the Naturalist says:
Male ducks are called drakes, female ducks are called ducks, and baby ducks are called ducklings.

Goose

Gander

Nora the Naturalist says:
Male geese are called ganders, female geese are called geese, and baby geese are called goslings.

16

Geese

Domesticated geese are gray geese and are kept for their meat, eggs, and soft down feathers. They have been farmed since ancient times.

Goslings

 Geese are herbivores and eat grasses, **rhizomes**, grain, and leaves.

 Geese are farmed all over the world. They are more expensive and difficult to raise than chickens.

 Average geese grow to around 30–35 inches (76–89 m) in length and weigh around 10 pounds (4.5 kg).

 Geese are members of the bird family. Domesticated geese cannot fly away because they are too heavy.

 A group of geese is called a "gaggle."

Horses

Horses were once a common sight on farms, where they pulled plows and carts. They are still used as working animals today on cattle ranches.

Mare

Foal

Nora the Naturalist says: Male horses are called stallions, female horses are called mares, and baby horses are called foals.

Horses eat grass.

Horses are found all over the world. They have even been taken on expeditions to the South Pole.

The size of horses varies by breed. Light riding horses range in height from 56–64 inches (142–163 cm) at the shoulder and can weigh from 840–1,200 pounds (380–550 kg).

Horses are mammals. They give birth to live young after an 11-months pregnancy.

There are more than 300 breeds of horses in the world today. They have many different uses.

Stallion

Donkeys and

Donkeys and mules are useful on farms. They are strong and can carry or pull heavy loads. Male donkeys are often kept to breed mules.

Donkey

Mules

Mule

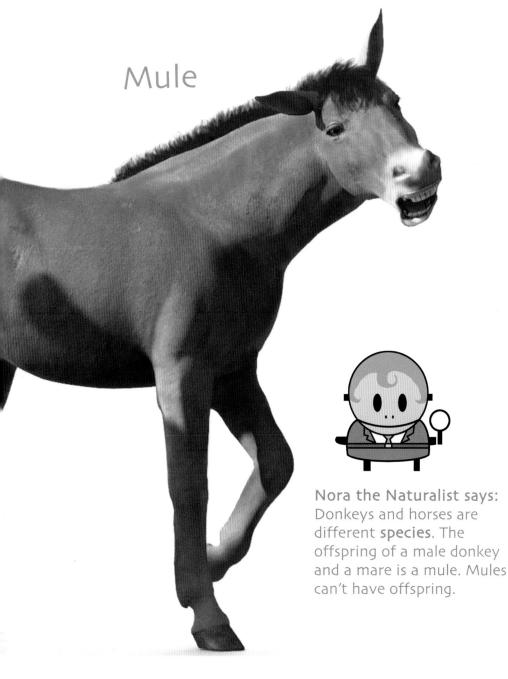

Nora the Naturalist says: Donkeys and horses are different **species**. The offspring of a male donkey and a mare is a mule. Mules can't have offspring.

Both donkeys and mules are herbivores.

Donkeys and mules are found all over the world.

A donkey measures from around 31 inches (79 cm) tall at the shoulder, and can weigh 180–1,060 pounds (80–480 kg). Mules come in a variety of shapes, sizes and colors, from under 50 pounds (20 kg) to over 1,000 pounds (500 kg).

Both donkeys and mules are mammals.

Donkeys are also called asses. A mule can carry up to 158 pounds (72 kg) and walk 16 miles (26 km) without resting.

Dog

Tom

Nora the Naturalist says:
Male cats are called toms, female cats are called mollies, and baby cats are called kittens. Male dogs are called dogs, female dogs are called bitches, and baby dogs are called puppies.

Dogs and cats are carnivores.

Dogs and Cats

Dogs are important working animals on farms. They help farmers and ranchers to herd sheep and cattle. Cats have a useful role, too, keeping **rodent** populations down.

Kitten

Molly

Dogs and cats are found on farms all over the world.

Dogs and cats are both mammals.

Sheepdogs like this Australian Koolie grow up to 24 inches (60 cm) at the shoulder and can weigh from 26–44 pounds (12–20 kg). Cats grow up to 12 inches (30 cm) at the shoulder and weigh around 10 pounds (22 kg).

Some types of sheepdog guard flocks of sheep from predators such as wolves.

23

Glossary

domesticated animals
Animals that have been changed over time by humans in order to meet their needs.

bullock
A castrated bull.

rhizome
Underground plant stem that looks like thick roots which stores food material.

rodents
Mammals that include mice, rats, and hamsters. Some species, like rats, are pests, eating seeds stored by people.

species
A group of living things capable of interbreeding.

steer
See bullock.

Index

cats 22–23
cattle 4–5
chickens 10–11

dogs 22–23
donkeys 20–21

ducks 14–15

geese 16–17
goats 12–13

horses 18–19

mules 20–21

pigs 8–9

sheep 6–7